This Notebook Belongs to :

Graphs of Sine, Cosine, and Tangent

The **sine function** has an up-down curve which repeats every 2π radians, or 360°.

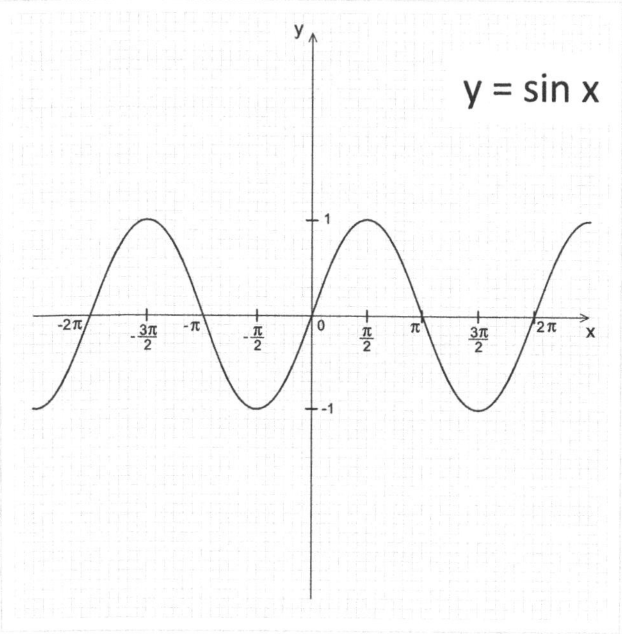

The **cosine function** has an up-down curve which repeats but it starts at 1 and heads down until π radians or 180°, and then heads up again.

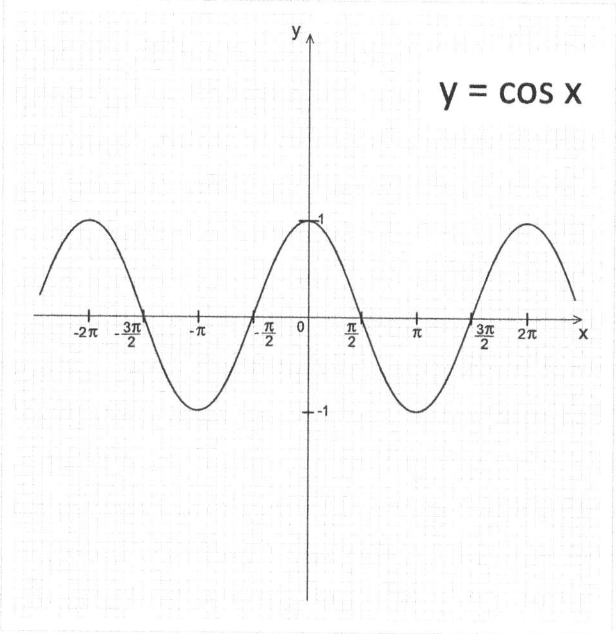

Notice the sine and cosine functions are like good friends. They follow each other, exactly $^{\pi}/_{2}$ radians or 90° apart.

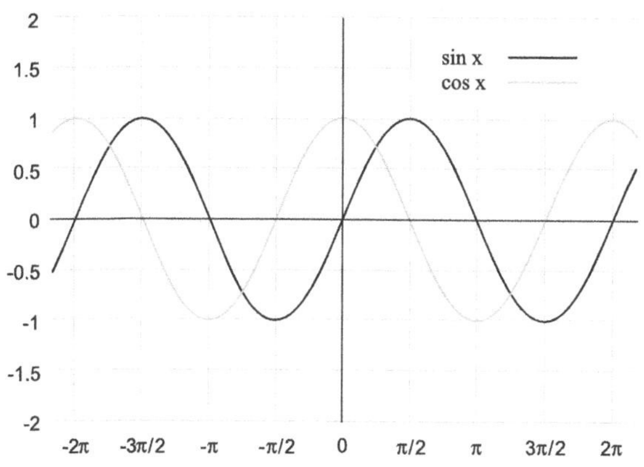

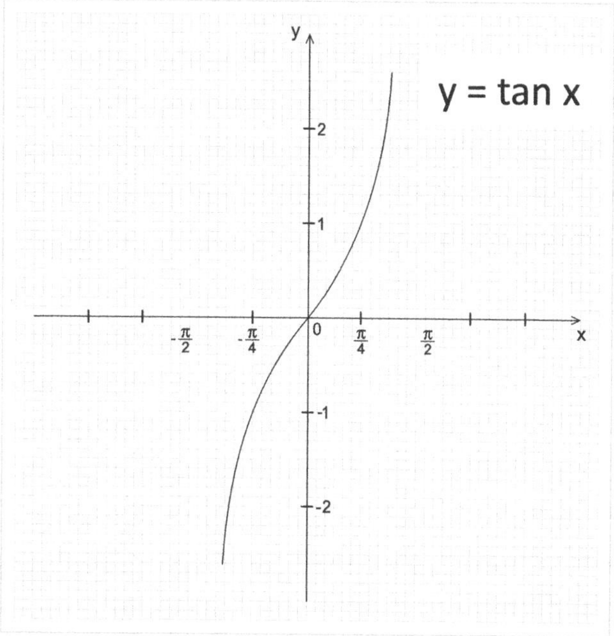

The **tangent function** goes between negative and positive infinity, crossing through 0, and at every π radians or 180°. At $^{\pi}/_{2}$ radians or 90°, and at $-^{\pi}/_{2}$ radians or $-90°$, $^{3\pi}/_{2}$ radians or 270°, etc., the function is officially undefined because it could be positive infinity or negative infinity.